Gabriela Lorenzon

Environmental impacts of implementing the Ludesa SHPP

Gabriela Lorenzon

Environmental impacts of implementing the Ludesa SHPP

Located in western Santa Catarina

ScienciaScripts

Imprint

Cover image: www.ingimage.com

This book is a translation from the original published under ISBN 978-613-9-70297-8.

Publisher:
Sciencia Scripts
is a trademark of
Dodo Books Indian Ocean Ltd. and OmniScriptum S.R.L publishing group

120 High Road, East Finchley, London, N2 9ED, United Kingdom
Str. Armeneasca 28/1, office 1, Chisinau MD-2012, Republic of Moldova, Europe
Printed at: see last page
ISBN: 978-620-8-18898-6

I dedicate this work mainly to my parents, Wilson and Cecilia (in memoriam), my boyfriend Adolfo and my uncle Luis Paulo, who helped me when I went out into the field, as well as to Forestry Engineer Aline Bortoncelo, for her help with the botanical classification of the tree species. Eternal gratitude to you and to my friends who believed in my research.

ACKNOWLEDGEMENT

I would firstly like to thank God for his strength and courage throughout this long journey. I would also like to thank all the teachers who accompanied me during my degree and especially my supervisor, Prof Janir de Oliveira Souza Júnior, for his guidance, dedication, patience and, above all, his friendship throughout the process.

I dedicate this, as well as all my other achievements, to my family and especially to my father Wilson Luis Antônio Lorenzon, for his trust, support and the opportunity to realise and complete another journey in my life. Thank you for teaching me all the ethical and moral values I know. I would like to say that being with you in this life is a great privilege. Without your understanding, help and trust, none of this would be possible.

To my beloved mum Cecilia (in memoriam), who unfortunately couldn't be present at this very happy moment in my life, but I couldn't fail to dedicate it to her, because if I'm here today, I owe many things to her and for her teachings and values. Thank you for everything! I miss you eternally!

To my boyfriend Adolfo Gasparetto Jr, for all his patience, understanding, affection, love and ineffable solidarity. Thank you for often helping me find solutions when they didn't seem to appear and for supporting me in all my decisions. You were the calm in difficult times, words are too few to thank you for your valuable help my love, without you none of this would be possible. My love for you is priceless.

To Forest Engineer Luis Paulo Lorenzon and Aline Bortoncelo for spending their weekends helping me with the forest inventory. Your help was extremely important in realising this dream. To Environmental and Sanitary Engineer Eduardo Miotello, for his immeasurable support and encouragement, for helping me put together the idea for the project, for his corrections, for making materials available and, above all, for his warm friendship.

To my friends and especially to Carla G. Gomes, you were my pillar throughout my degree, thank you for always being by my side, without you nothing would have been possible. I would also like to thank the Gasparetto family, who have always welcomed me so well and have always been willing to help me with anything I needed, thank you!

"A little perfume always stays in the hands of those who offer flowers"

(Chinese proverb)

SUMMARY

Economic activity, together with the growth in electricity demand and the availability of water resources, has led to hydroelectric power becoming Brazil's biggest energy potential. Small hydroelectric plants currently account for 2.9% of the electricity generated in Brazil, representing around 4.67% of their capacity in the state of Santa Catarina. What is worrying is the proportionally greater need for electricity, as well as the lack of socio-economic planning and the fact that people are not aware of this scenario. The aim of this research is to analyse the effects of the environmental impacts of the Ludesa Small Hydroelectric Power Station located in the municipality of São Domingos, in the western region of the state of Santa Catarina. To this end, the changes to the biotic environment (fauna and flora) were assessed before and after the installation of the project, analysing their alterations. Fauna data from the development area was collected through direct and indirect observations of the animals. The methodology included interviews, trace analysis and the installation of galvanised traps. Phytosociological data was collected using four 30x30m plots drawn within a radius of 500 metres from the Ludesa SHPP dam axis and by comparing aerial photogrammetric images before and after the project was installed. The data was interpreted using tables, graphs, photographic records and descriptive analyses. Sixteen mammal species and 27 different botanical species were identified.

Keywords: PCH, Fauna, Flora.

SUMMARY

CHAPTER 1	**6**
CHAPTER 2	**8**
CHAPTER 3	**19**
CHAPTER 4	**30**
CHAPTER 5	**43**
CHAPTER 6	**45**
CHAPTER 7	**49**

CHAPTER 1

INTRODUCTION

Since the Industrial Revolution in the 18th century, the exploitation of natural resources has become increasingly frequent, so that they were seen as inexhaustible. Entrepreneurs cared only about the economic viability of their projects. It was only with the occurrence of major natural phenomena and disasters that environmental concerns became relevant and it was only in 1972 with the Stockholm Conference that environmental issues began to be considered during the development of economic projects (CARVALHO, 2010).

The construction of energy projects has been growing rapidly in Brazil, due to its physical characteristics and the presence of flowing rivers, which have made the country the third largest hydroelectric potential in the world. According to data from the World Bank, hydroelectric power stations accounted for 85 per cent of the installed capacity of Brazil's energy system in 2008 (CARVALHO, 2010).

For many years it has been said that hydroelectric power does not cause any significant environmental impact on the environment, but recent studies show that this type of generation, once considered "clean energy", has been causing serious impacts from its construction to its operation (BATISTA, 2012).

Hydroelectric power plants don't always achieve their objective and sometimes end up producing less electricity than they were designed to, causing many companies to start new projects before the environmental problems of the plants they have built have even been resolved (MENDES, 2005).

Some of the environmental impacts generated in the implementation phase up to the operation of these projects are the hydrogeological alterations to their flow, the exaggerated increase in hydraulic detention time, the alteration in the flow rate, the change in the aquatic environment from lotic to lentic, the change in temperature and relative humidity and the loss of biodiversity (BRITO, 2012).

Despite all the environmental impacts generated by hydroelectric projects, hydroelectric plants are still seen as a more viable alternative due to their low production costs, after all, the energy generated comes from renewable and available resources (BATISTA, 2012).

In Brazil, where a large part of its energy comes from electricity, it has become necessary to develop environmental impact studies - EIA, a management tool that can minimise the environmental degradation caused by small hydroelectric plants, and the development of such studies is indispensable for their environmental licensing. It was found that 50% of the projects have no environmental licensing problems, but 43.3% have environmental problems and 5.7% have various problems (FACURI, 2004).

Population growth together with the proportionally greater need for electricity, as well as the lack of planning in the socio-economic model, contributes to the serious environmental problems mentioned above, considering that people's lack of awareness could culminate in significant future adversities in the global ecosystem of Planet Earth (FACURI, 2004).

The aim of this work is to analyse the effects of the environmental impacts of the Ludesa Small Hydroelectric Power Station. The aspects relating to the biotic environment (fauna and flora) were assessed before and after the installation of the project, verifying possible alterations.

CHAPTER 2

THEORETICAL REVIEW

2.1 HISTORY AND CHARACTERISATION OF THE PROJECT

The Ludesa Small Hydroelectric Power Station was granted a licence by ANEEL Resolution 705 of 2002, in which it submitted its Preliminary Environmental Licence (LAP), issued by the Environmental Foundation (FATMA). After the Environmental Installation Licence (LAI) was submitted, ANEEL Order No. 116 of 17 March 2003 was issued, in which the Superintendence of Hydraulic Potential Management (SPH) approved the plant's basic project (NOBREGA, 2012).

ANEEL issued Order No. 2,401 on 31 July 2007, supported by an Environmental Operating Licence (LAO), which allowed the Superintendence for the Supervision of Generation Services to authorise the operation of the project (NOBREGA, 2012).

The Ludesa Small Hydroelectric Power Station is located 200 kilometres from the mouth of the Chapecó River, and the dam is located in the municipalities of São Domingos (right bank) and Ipuaçu (left bank), whose artificial reservoir also covers the municipality of Abelardo Luz, noting that the water intake, penstock and powerhouse are located in Ipuaçu, in the west of Santa Catarina. The dam is 29 metres high and 692 metres long, and the spillway is 178.8 metres long, as shown in Figure 01. The flooded area is 8.17 km^2 at its maximum normal level (RIPPEL, 2013).

The powerhouse has two generating units, each with 15 MW of installed power, for a total of 30 MW, so that the two turbines take advantage of a gross drop of 46.2 metres, as shown in Figure 02. At the substation, the voltage is raised to 138 kV and the energy generated is transmitted via a single-circuit transmission line with a length of 24 kilometres, of which 22 kilometres are in Santa Catarina and 02 kilometres in the state of Paraná, to its receiving substation in Clevelândia, Paraná, as part of the national electricity circuit, managed by the National Electricity System Operator - ONS. (RIPPEL, 2013).

Figure 1 - Ludesa Small Hydroelectric Power Plant.

Source: Rippel, 2013

Figure 2 - Ludesa SHPP power house.

Source: Rippel, 2013.

2. 2LOCAL CHARACTERISATION

The Ludesa Small Hydroelectric Plant is located between the municipalities of São Domingos, Abelardo Luz and Ipuaçu, in Santa Catarina, as shown in Figure 03 (ABDO, 2002).

It is located on the Chapecó River, part of the Uruguay River Basin, at geographical coordinates: Latitude 26°34'38" S and Longitude 52°29'46" W. (ABDO, 2002).

The basic project and engineering studies for the Ludesa SHPP were carried out by Ludesa Energética S/A (ABDO, 2002).

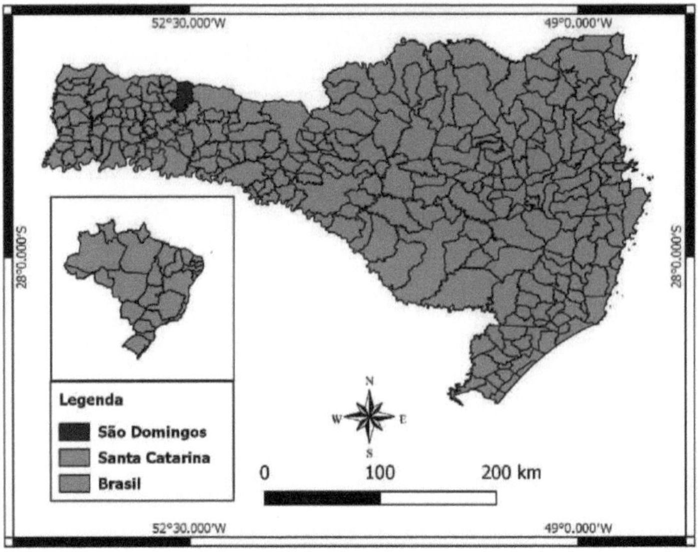

Figure 3- Location of the Municipality of São Domingos in the State of Santa Catarina. Source: Prepared by the author; QGIS, 2016.

Access to the Ludesa SHPP is via the SC 480 highway, in the town of Nossa Senhora dos Navegantes, in the municipality of São Domingos, where the dam is located approximately 500 metres away (ABDO, 2002).

2.2.1 Climate and weather conditions

The Ludesa SHPP is located in the humid mesothermal climate type Cfb according to Koppen, which corresponds to a subtropical climate with mild summers and harsh winters with no defined dry season. It corresponds to a climate with average temperatures of between 18°C and 22°C. The Cfb climate is influenced by latitude and altitude, as well as by air masses that contribute to the region's cold in winter. The rainfall regime is of the isoigneous type, i.e. well-defined rainfall throughout the year (CHEREM, 2008).

2.1.1 Geology and Geomorphology

The regional geological cover dates back to the Cretaceous period and is made up of basaltic rocks from the Serra Geral formation. In this respect, effusive volcanic rocks are described, represented by effusions which cover around 50% of the state of Santa Catarina (CRUZEIRO, 1991).

These spills are present in several Brazilian states and part of the southern region of Brazil. In the state of Santa Catarina, this area corresponds to around 52% of the territory. This type of rock has structures such as basalt, phonobasalt, realites and dacites (WELTER, 2006).

In the west of Santa Catarina, around 80 per cent of the area is covered by the Serra Geral Formation. In terms of natural resources, the ones that stand out are: natural water, amethyst, basalt, red clay, copper and manganese (CRUZEIRO, 1991).

Basalts are basically made up of plagioclase, pyroxene, magnetite and secondary minerals. Plagioclase is the most abundant component and its stability is very low, so it ends up turning into clay minerals with a high release of iron oxides, causing what is known as weathering. Basalt soils therefore have a clayey texture (CHEREM, 2008).

2.1.2 Pedological aspects

The soil classes found in the region are Latossolo, Cambissolo and Argissolo (SDR, 2003).

The Ludesa SHPP area has five dominant soil classes: Cambissolo háplico, Gleissolo melânico, Latossolo vermelho, Neossolo litólico and Nitossolo vermelho (CHEREM, 2008).

Cambisols are soils with a shallower depth (from 0.5 to 1.5 metres). They occupy approximately 52% of the state of Santa Catarina, are located on undulating terrain and have varied natural fertility. Those with high fertility are used for growing corn, beans and soya, while those with low fertility are used for reforestation and pasture (CRUZEIRO, 1991).

Gleissols occupy flat terrain and are poorly drained. They are clayey, plastic and sticky. Chemically, they have low nutrient availability and high acidity (LEPSCH, 2010).

Latosols are made up of deeper, porous and well-drained soils. They are found on undulating terrain and usually have low natural fertility (CRUZEIRO, 1991).

Neosols are well-drained soils with intense stoniness on the surface and in the profile. They are morphologically young soils and occupy strongly undulating reliefs (LEPSCH, 2010).

Nitosols are rich in iron and aluminium derived from basic effusive rocks. They are very deep, purplish-coloured soils with a loamy texture, well porous and drained. They are located on very undulating and mountainous terrain (LEPSCH, 2010).

2.1.3 Water Resources

The Ludesa Small Hydroelectric Power Station area is located in the Uruguay River basin, whose hydrographic region is characterised by regular rainfall well distributed throughout the year, with a monthly total of more than 60 mm (MMA, 2006).

The project is located in Hydrographic Region 2 - Midwest. This hydrological unit is formed by the Chapecó and Irani river basins. The Chapecó River has an average flow of 263 m^3 and a drainage area of 8,190 km^2 , representing the main watercourse in RH-2, draining 30 municipalities including São Domingos, Abelardo Luz and Ipuaçu (EAS - PCH Ludesa, 2006).

The rivers in this basin, such as the Chapecó River, have numerous waterfalls, which represent a wealth of hydroelectric potential for the western region of Santa

Catarina (EAS - PCH Ludesa, 2006).

2.3 ASPECTS OF FAUNA IN THE STUDY REGION

Some groups of animals can be used as indicators of environmental quality, and characteristics such as absence or presence, abundance and reproductive success can represent the environmental quality of a given environment (CARVALHO, 2008).

The state of Santa Catarina has a rich fauna and, according to the Environment Foundation, is represented by 169 species described in the national register of Brazilian fauna species (SILVA, 2011).

Wild mammals are difficult to see and, consequently, describe because their habits are often discreet and nocturnal, and their identification is made by signs such as footprints, faeces, etc. (SILVA, 2011).

Environmental problems such as deforestation and forest fragmentation are serious in western Santa Catarina, where knowledge about the mastofauna occurring in the study region is very weak and scientific knowledge is found in few academic materials (CHEREM, 2008).

In the study area, there are some species of mammals threatened with extinction according to CONSEMA Resolution 002/2011, such as the Preá *(Cavia intermédia),* Jaguatirica *(Leopardus pardalis),* Cateto *(Pecari tajacu),* Anta *(Tapirus terrestris),* Paca *(Cuniculus paca)* and some species of bat (FATMA, 2012).

Currently, around 7,150 species of scaled reptiles have been identified, including 2,700 species of snakes, 4,000 species of lizards and 140 species of amphisbaenians. Around 641 species of reptiles have been identified in Brazil, and this number could increase (CHEREM, 2008).

In Santa Catarina, some species of endangered reptiles have been identified in accordance with CONSEMA Resolution No. 002 of December 2011, namely the Green Muçurana *(Cletia plúmbea), the* field lizard *(Cnemidophorus vacariensis)* and the barb-tailed cagado *(Prhynops williamsí).* Not many endangered reptile species are found in the Ludesa SHPP region (FATMA, 2012).

Brazil is home to a significant amount of the world's amphibian biodiversity, in which the Atlantic Forest represents the biome with the greatest diversity of anurans in Brazil and the greatest number of endangered species are found in it (GONSALES, 2009).

For the state of Santa Catarina, 110 species of amphibians have been recorded, 12 unidentified species and 22 species with probable occurrence. These species correspond to 17% of the species found in Brazil and 35% of the species found in the Atlantic Rainforest (IAP, 2002).

The Amphibia class has around 6,091 species, which are divided into three orders: Anura (frogs, toads and tree frogs), Caudata (salamanders) and Gymnophiona (blind snakes). Around 97 species of amphibians have been described in recent years in the country, suggesting that the diversity is even greater than currently described. However, environmental degradation is gradually increasing and many species are disappearing without even being known about (CHEREM, 2008).

For the state of Santa Catarina, some species of amphibians are identified as endangered according to CONSEMA Resolution 002/2011, namely the Brown Perereca *(Aplastodiscus cochranae)* and the Glass Frog *(Vitreorana uranoscopa)*. Not many species of endangered amphibians are found in the Ludesa SHPP area (FATMA, 2012).

Birds have characteristics that make them ideal for describing the conservation of a given environment. They occupy all altitudes and longitudes of the planet. They are considered to be the easiest taxon to study. This is due to the ease with which they can be identified in the field, through direct (visual) or auditory observation with the aid of field guides. Birds are also considered to be excellent bioindicators, as they occupy the most diverse ecological niches and food guilds and can be used to assess a given area and its state of conservation (CHEREM, 2008).

Brazil has 1,801 species of birds occurring on its mainland. For the Atlantic Forest, one of the biomes with the greatest endemism, around 690 species of birds have been described, of which around 200 species are endemic and 150 species are in the threatened category, mainly due to habitat destruction, so that currently less than 5%

of this forest's vegetation cover remains (CHEREM, 2008).

For the state of Santa Catarina, a number of endangered bird species have been identified according to CONSEMA Resolution 002/2011, including the Curió *(Sporophila angolensis),* Araçari-castanho *(Pteroglossus castanois),* Pintadinho *(Drymophila squamatd)* and the Royal Hawk *(Harpia harpyja').* Not many endangered bird species are found in the Ludesa SHPP area (FATMA, 2012).

There are around 28,000 species of fish in the world, of which around 40 per cent are freshwater. The freshwater fauna is the most diverse on the planet and 2,400 species have already been identified and catalogued, more than half of which belong to the Amazon River basin (CHEREM, 2008).

The Uruguay River Basin has around 250 species of fish, the most predominant of which are the Characiform and Siluriform orders, which are able to adapt to different environments and currently occupy the most diverse ecological niches (CHEREM, 2008).

Ichthyofauna species that suffer from the changes caused by hydroelectric projects are subdivided into two groups: those composed of rheophilic species, which occur in running water, and those composed of species that inhabit lithic environments, with deep areas and flooded regions (ABILHOA, 2003).

The Chapecó River is home to fish species such as Cascudos *(Hypostomu* spp.), lambaris *(Astyanax* spp. and *Bryconamericus* spp.), mandi *(Pimelodus* spp.), ladybirds *(Crenicichla* spp.) and dogfish *(Oligosarcus* spp.). This identification is due to Ichthyofauna studies for Environmental Reports for the construction of Hydroelectric Power Plants, in which they appear in reasonable numbers along the course of the Chapecó River (CHEREM, 2008).

2.4 PHYTOPHYSIOGNOMY OF THE FLORA IN THE STUDY REGION

The state of Santa Catarina is part of the Atlantic Forest biome, in which around 81 per cent of its territory is occupied by forests, 24 per cent of which is covered by native forests. The Atlantic Forest has been considered a "hotspot", defined as a biome of high biological biodiversity worldwide (REZENDE, 2013).

The state of Santa Catarina has three types of forest: Dense Ombrophilous Forest, Mixed Ombrophilous Forest and Stationary Deciduous Forest (VIBRANS, 2015).

The vegetation around the Ludesa SHPP is part of the Mixed Ombrophilous Forest (FOM), which extends throughout the basin of the River Uruguay and its tributaries. This phytophysiognomic vegetation is due to the climate of rainy and dry seasons, with marked thermal variation (CHEREM, 2008).

The Mixed Ombrophilous Forest or Araucaria Forest generally occurs at altitudes of over 500 metres, with the Araucaria as the main species, dominating the forest canopy, i.e. it has a continuous cover formed by the treetops (VIBRANZ, 2013).

The vegetation of the region where Araucaria occurs is not a homogeneous and continuous formation, but rather is made up of multiple associations and groupings with various successional stages (KLEIN, 1984).

The most frequent emerging species at the Ludesa SHPP are the grápia *(Apuleia leiocarpà),* angico-vermelho *(Parapiptadenia rígida),* guajuvira *(Patagonula americana),* Maria preta *(Diatenopterix sorbifolia),* canafistula *(Peltophorum dubium)* and cedar *(Cedrela fissilis*), among others (CHEREM, 2008).

2.5 ENVIRONMENTAL LICENSING

Environmental Licensing is an instrument defined by Federal Law No. 6938, of 31 August 1981, which was founded by the National Environmental Policy, with the aim of promoting prior control over the construction of projects that use natural resources and are considered potentially polluting (MMA, 2009).

The environmental licensing process *is* governed by legal norms established by CONAMA Resolution 001 of 23 January 1986, which establishes guidelines for the preparation of Environmental Impact Studies (EIA) and Environmental Impact Reports (RIMA) in the environmental licensing process (MMA, 2009).

For the state of Santa Catarina, all entrepreneurs must comply with CONSEMA

Resolution 13/2012, together with FATMA's Normative Instruction - IN 44, which deals with the general and specific guidelines for the environmental licensing of hydroelectric projects, as well as the list of necessary documentation, These include the Simplified Environmental Study (EAS), the Environmental Compliance Study (ECA), the Environmental Control Plan (PCA), the Detailed Environmental Programme Report (RDPA), the Degraded Area Recovery Plan (PRAD) and the Emergency Action Plan (PAE), depending on the characteristics of each project. (FATMA, 2012).

The relevant environmental licensing instruments are the Preliminary Environmental Licence for locational approval based on the relevant environmental studies, the Installation Environmental Licence to authorise the implementation of the project based on the detailed projects, and the Operating Environmental Licence to authorise the operation of the project on verification of compliance with the previously approved projects and on the condition of compliance with environmental control monitoring plans, issued by the Environment Foundation (FATMA, 2012).

It should be noted that the Environmental Operating Licence is usually valid for four years, so that with each renewal the enterprise must submit a technical environmental report proving compliance with the requirements and conditions established in the previous Environmental Operating Licence (FATMA, 2012).

2.6 ENVIRONMENTAL IMPACT ASSESSMENT

The environmental impacts caused by hydroelectric projects are a matter of controversy in the discussion on sustainable development, with SHPs causing negative impacts on the environment. The big question *is* knowing the real scale of these impacts and how they can be mitigated (BONSOR, 2016).

The impacts resulting from the construction of a SHPP should be well reported and documented, and are related to the size, geographical location, volume and retention of the reservoir. All these changes can have direct and indirect effects, transforming the biogeophysical, economic and social conditions of a given area

(NILTON, 2009).

The ecological integrity of an aquatic ecosystem is linked to its water flow regime. The construction of PCTTs alters the natural dynamics of this ecosystem, and its modifications can have three effects: the physical, chemical and biological characteristics of the watercourse, its primary productivity and the communities of consumers (LATINI, 2016).

The environmental impacts caused by dams are well known and can lead to global, regional and local impacts. Global impacts relate to the influence the dam can have on the climate due to the suppression of vegetation and the filling of the reservoir. Local impacts are due to the environmental consequences that occur around the dam caused by blocking the flow of sediment, altering the hydro-geological cycle, preventing fish migration, reducing oxygen levels upstream and downstream of the dam, fragmentation of the aquatic ecosystem and loss of biodiversity (LATINI, 2016).

The installation and operation of SHPs alters the physical characteristics of a river, changing its frequency from lotic (flow with greater velocity) to lentic (flow with less velocity) and, consequently, altering the structure of existing habitats along its course. These alterations can cause changes in its biota, altering the structure of local populations and communities (LATINI, 2016).

CHAPTER 3

METHODOLOGY

This study was carried out over a period of 09 months, from March to November 2017, where surveys were carried out in the area of the development to collect data for the fauna and flora survey in the area of the development.

3.1 FAUNA SURVEY IN THE STUDY REGION

In order to assess the fauna in the study area, a methodology was adopted consisting of direct and indirect observation of the species. By applying this methodology, the qualitative values of the site were presented, showing the degree of occurrence of the species and the existing impacts of the construction work. An area of coverage was considered for the assessment, drawn within a radius of 1,000 metres from the dam and the Area of Direct Influence - AID of the Ludesa SHPP.

3.1.1 Method Used Through Direct Observation

The direct observation method involves seeing the animal *in* real time, where this can happen during *on-site* visits to the development area. For direct observation, galvanised traps, *pitfall* traps and active search were used. The methodology used consisted of walking slowly along the edges of the target and along existing trails. This procedure was carried out carefully, in silence and with a maximum of two people per sample area, starting at dawn and late afternoon, as these are the times when the species use the environments to move around and look for food.

3.1.1.1 Galvanised traps

According to Branco Jr (2006), Sherman and Tomahawh type cages or galvanised wire grid cages are used to capture small mammals. The location of the traps should be determined by analysing the behaviour of the species, as some species are only caught near rivers and others in denser vegetation. The choice of bait is also important as some mammals have a predilection for certain types of bait.

These baited traps were placed at three strategic points in the woodland in the Ludesa SHPP's AID, as shown in Figure 04, indicating the location of the points where the traps were installed.

Figure 4 - Location of the trap installation points in the study area.

Source: Adapted from Google Earth, 2016.

Table 01 shows the geographical coordinates Latitude, Longitude of the trap installation points in the development area.

Table 1 - Geographical coordinates of the galvanised traps installed in the development area

Trap location	Latitude	Longitude
Point 01	26°34'48.99"	52°29' 15.41"
Point 02	26°34'57.98"	52°29' 19.26"
Point 03	26°34'22.00"	52°29'56.91"

Source: Google Earth, 2016.

The galvanised traps were set up in the late afternoon and inspected at dawn. Pieces of bacon and chicken were used as bait because of their attractive smell. The traps were placed close to the reservoir and at passageways through the forest.

The traps installed in the development area can be seen in Figure 05.

Figure 5 - A) Sherman model trap installed in the study area. B) Tomahawh model trap installed in the study area.

Source: The author, 2017.

3.1.1.2 Interception and Drop Traps

The interception and fall traps, also known as pitfall traps, consisted of a set of buckets which were buried with the entrance open. There were three buckets interconnected by canvas fences used to guide the animal to the bucket. They are widely used to capture smaller animals such as reptiles and amphibians.

This trap was installed at one point in the development's Permanent Preservation Area, as shown in figure 07, indicating the location of the point where the trap was installed. On each visit it was reactivated and checked at dawn, midday and late afternoon.

Figure 6 - Location of the Interception and Drop Trap installation point.

Source: Adapted from Google Earth, 2016.

Table 02 shows the geographical coordinates Latitude, Longitude of the trap installation point in the development area.

Table 2 - Geographical coordinates of the interception and drop traps installed in the development area

Trap location	Latitude	Longitude
Point 01	26°34'51.38"	52°28'44.55"

Source: Google Earth, 2016.

The interception and fall trap installed in the study area can be seen in figure 07.

Figure 7 - Interception and Drop Trap installed in the development area. Source: The Author, 2017.

3.1.1. 3Active Search

It consists of searching all environments where anurans, lizards, snakes and arthropods may be present, such as tree hollows, leaf litter, holes, burrows in the ground and under rocks. Active searches are being carried out throughout the Ludesa SHPP's Area of Direct Influence.

3.1.2 Method Used Through Indirect Observation

The indirect observation method involves seeing the traces found at the study site, such as footprints, faeces, fur, grunts and vocalisations. With this method it is possible to prove the presence of individuals without visualising them. Footprint traps and interviews with neighbouring residents can be used for indirect observation.

3.1.2.1Footprints and faeces

It consists of analysing the animal footprints and faeces found in the AID, where they were left when moving between forest fragments. The method can be carried out through direct observation and by installing footprint traps, which consist of distributing a parcel of sand with bait such as banana slices and bacon in areas of the forest. The footprint traps were placed at five strategic points in the woodland in the Ludesa SHPP's AID, as shown in Figure 08, indicating the location of the points where the footprint traps were installed.

Figure 8 - Location of the footprint trap installation points.

Source: Adapted from Google Earth, 2016.

Table 03 shows the geographical coordinates Latitude, Longitude of the trap installation points in the development area.

Table 3 - Geographical coordinates of the footprint traps installed in the development area

Trap location	Latitude	Longitude
Point 01	26°34'48.41"	52°29'33.63"
Point 02	26°34'59.67"	52°29'14.30"
Point 03	26°34'46.10"	52°29'34.61"
Point 04	26°34'33.48"	52°30'00.27"
Point 05	26°34'50.59"	52°29'18.37"

Source: Google Earth, 2016.

The footprint trap installed in the development area can be seen in Figure 09.

Figure 9 - Footprint trap installed in the study area.

Source: The author, 2017.

3.1.2.2 Interviews

Interviews were held with residents near the Ludesa SHPP's IDA, where questions were asked about the region's fauna.

3.2 FLORA SURVEY IN THE STUDY REGION

3.2.1 Use of Sampling and *On-site* Inventory

The Area of Direct Influence - AID of PCH Ludesa was considered to be the place where simple random sampling was carried out on the remaining vegetation, considering a percentage of approximately 2% of the vegetation within a radius of 500 metres from the axis of the PCH Ludesa dam, where four 30m X 30m plots were built, totalling 3. 600m , in order to know the phytosociological parameters.600m^2 , in order to find out the phytosociological parameters Density (Absolute and Relative), Dominance (Absolute and Relative), Frequency (Absolute and Relative), Importance Value Index, Cover Value Index and Species/Area Curve.

3.2.1.1 Density

Density expresses the participation of different species within the vegetation context of a given area. It can be calculated in two ways: absolute density, which shows the number of individuals of a given species per unit area (ha), and relative density, which shows the participation of species in relation to the total number of individuals (IAP, 2002).

Absolute and relative densities are obtained by calculation using the following expression:

$$DA = \frac{n}{Área}$$

Expression 1 - Absolute Density Formula

Where:
DA = Absolute Density
n = number of individuals of a given species

$$DR = \left(\frac{n}{N}\right) * 100$$

Expression 2 - Relative Density Formula

Where:
RD = Relative Density
N = Total number of individuals
n = Number of individuals of a given species.

3.2.1. 2Frequency

Frequency represents the distribution of each species in the area. It can be shown in two ways: Absolute Frequency indicates the percentage of plots with a particular species and Relative Frequency indicates the ratio of the absolute frequency of each species to the sum of the absolute frequencies of all species (IAP, 2002).

Absolute and relative frequencies are calculated using the following expression:

$$FA = \frac{pi}{P} * 100$$

Expression 3 - Absolute Frequency Formula.

Where:

AF = Absolute Frequency

Pi = Number of plots with occurrence of species i.

P = Total number of plots in the sample.

$$FR = \frac{FAi}{\Sigma \mathrm{FA}} * 100$$

Expression 4 - Relative Frequency Formula.

Where:

RF = Relative Frequency

FAi = Absolute Frequency of a given species

ΣFA = Sum of the absolute frequencies of all the species sampled.

3.2.1. 3 Dominance

Dominance refers to the rate at which individuals of a species occupy the environment in a given area. It can be demonstrated in two ways: Absolute Dominance, which represents the basal area of a given species per area, and Relative Dominance, which represents the ratio of the absolute dominance of a given species to the sum of the absolute dominance of all species (IAP, 2000).

Absolute and relative dominance are calculated using the following expression:

$$DoA = \frac{gi}{Área\ (ha)}$$

Expression 5 - Absolute Dominance Formula.

Where:

DoA = Absolute Dominance

gi = Total basal area of a given species

$$DoR = \left(\frac{gi}{G}\right) * 100$$

Expression 6 - Relative Dominance Formula

Where:

DoR = Relative Dominance

gi = Total basal area of a given species

G = Basal area of all species sampled

3.2.1.4 Importance Value Index

And the index that indicates the importance of species in the community, bringing together the criteria from analysing the three parameters DR, FR, DoR (IAP, 2002).

The Importance Value index is calculated using the following expression:

$$IVI = DR + FR + DoR$$

Expression 7 - Importance value index formula

Where:

IVI = Importance Value Index

RD = Relative Density

RF = Relative Frequency

DoR = Relative Dominance

3.2.2 Use of Satellite Images

In order to dimension the areas associated with the Ludesa SHPP, a visual analysis was made of the satellite images available on the Google Earth Pro application. Thus, after comparatively analysing the images available in 2004 and 2017, i.e. before and after the installation of the project, the variation in forest cover between the periods was assessed, as well as whether the legislation in force regarding Permanent

Preservation Areas (PPAs) had been complied with.

The study area was delimited from a radius of 1,000 metres from the axis of the Ludesa Small Hydroelectric Power Plant dam. The remaining native vegetation areas were delimited there, as well as the Permanent Preservation Areas existing before and after the project.

CHAPTER 4

RESULTS AND DISCUSSION

4.1FAUNA OF THE STUDY REGION

4.1.1 Method Used through direct observation

The galvanised traps and the interception and drop traps were set up in the study area on 31 March. The galvanised traps were set up again at four different times, on 15 April, 29 April, 10 June, 17 June and 24 June. The traps were found to be empty in the days following installation, often being turned over and in other environments.

According to Oliveira (2011), mammals are among the most difficult animals to study because they occur in low densities, have a more heterogeneous distribution, cryptic behaviour and are difficult to locate in dense vegetation. For this reason, studies aimed at estimating the abundance and occurrence of mammals are more likely to be based on indirect observation through footprints, faeces, fur, burrows and so on.

4.1.2 Method used through indirect observation

The indirect observation method proved to be efficient in verifying the occurrence of endemic species at the study site, which were recorded and confirmed using their traces. This method made it possible to verify the presence of individuals without visualising them.

The occurrence of two mammal species was recorded: the Catingueiro deer *(Mazama gouazoubird)* and the capybara *(Hydrochoerus hydrochaeris)* through footprint traces. The species were identified with the help of the Paraná Fauna Tracks Manual.

The characteristics of the tracks of the Catingueiro Deer *(Mazama gouazoubirà)* (figure 10) are of the unguligrade type, where its markings are long laterally and the impression of each digit is widened on the posterior margin, narrowing on the anterior margin which ends in a pointed nail (MORO-RIOS, 2008), as can be seen in figure 11.

Figura 10 - Footprint of Catingueiro deer *(Mazama gouazoubirà)* found in the study area.

Source: The author, 2017.

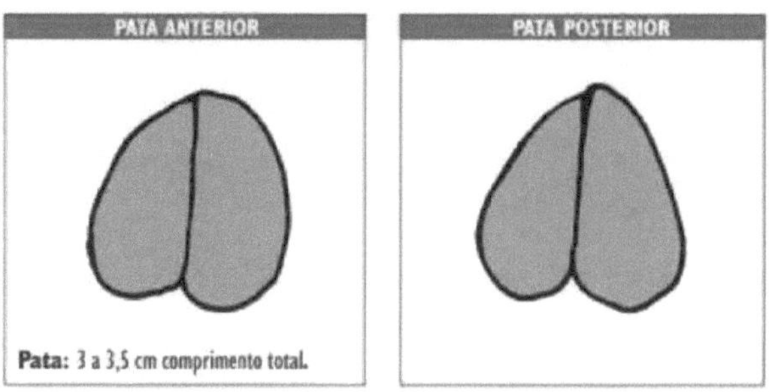

Figura 11 - Illustration to identify the footprint of the Catingueiro Deer *(Mazama gouazobirà).*

Source: Moro-Rios, 2008

The characteristics of the capybara*'s (Hydrochoerus hydrochaeris)* footprints, shown in figure 12, are plantigrade, with four digits on the hind feet and three digits distributed radially on the forelegs. The footprints of the forelegs are made up of the second, third and fourth digits, while the hind feet are larger, showing three elongated toes, the median of which is larger (MORO-RIOS, 2008), as can be seen in figure 13.

Figure 12 - Capybara *(Hydrochoerus hydrochaeris)* footprint found in the study area.

Source: The author, 2017.

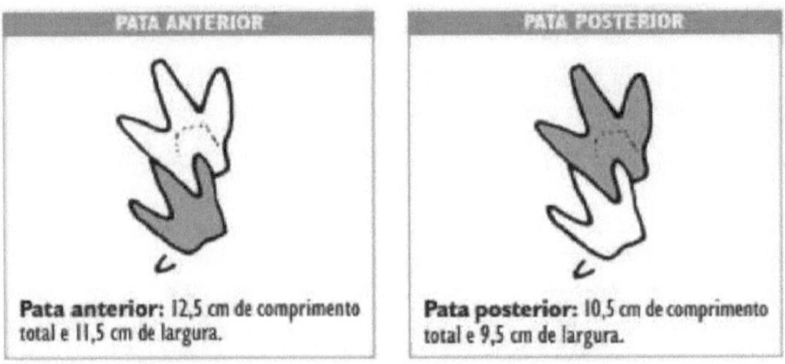

Figure 13 - Illustration to identify the footprint of the Capybara *(Hydrochoerus hydrochaeris).*

Source: Moro-Rios, 2008.

During walks in the area of the development, it was possible to verify the presence of two species of mammal through traces of faeces. The Graxaim *(Cerdocyon thous),* figure 14, and the Capybara *(Hydrochoerus hydrochaeris*), figure 15, were recorded.

According to Oliveira (2011), the Graxaim *(Cerdocyon thou)* faeces are identified by analysing the morphological and microstructural characteristics of the hairs found in the faeces, looking at the colour of the hairs, the cuticular pattern and the medullary pattern. The colour of the fur is white and black interspersed from the base to the apex, with the base being white and the apex black. The cuticular pattern is characterised by pavement-like, transverse and wavy scales with some irregular scales and smooth edges. The medullary pattern shows juxtaposed multiseriate cells with an anisocellular shape and an intact margin.

Figure 14 - *Cerdocyon thous* faeces.

Source: The author, 2017.

Figure 15 - Capybara faeces *(Hydrochoerus hydrochaeris).*

Source: The author, 2017.

According to Moro-Rios (2008), footprints and faeces are an easily accessible form of trace, which makes their use an efficient alternative for identifying endemic species in the region studied. According to the data, information such as

species abundance, habitat indicators, feeding habits and indicating the presence of other species by analysing the faeces of carnivorous species.

The interviewees reported the occurrence of the following species: Mulita armadillo *(Dasypus hybridus),* Catingueiro deer *(Mazama gouazoubird),* Quati *(Nasua nasuã),* White-eared opossum *(Didelphis albiventris),* Otter *(Lontra longicaudis),* Capybara *(Hydrochaeris hydrochaeris),* European hare *(Lepus europaeus),* Graxaim or Cachorro- do-mato *(Cerdocyon thous),* Preá *(Cavia sp.), Procyon cancrivorus, Sapajus nigritus* and *Leopardus wiedii.*

Twelve mammal species were identified, as shown in Table 4, grouped into 7 orders and 10 families. The most representative in terms of individuals is the order Carnivora with 42% of the species recorded and the most representative families are Procyonidae and Felidae with 17% of the species recorded. As can be seen in graphs 01 and 02.

Table 4 - List of Fauna Species Identified in the Ludesa Small Hydroelectric Power Plant Study Region:

Species	Popular Name	Identification							FAUNA
		AR	EN	VE	RB	VD	FE	PE	
Dasypus hybridus	Mulita armadillo		X	X	X				
Mazana sp.	Deer	X	X	X	X			X	
Nasua nasua	White-eared possum		X		X				
Otter longicadus	Otter		X		X				
Hydrochaens hydrochaeris	Capybara	X	X	X	X		X	X	
Lepus europaeus	European hare		X		X				
Cerdocyibthous	Graxaim (bush dog)		X	X	X		X		
Cavia sp.	Preá		X		X				
Procyon cancrivorus	Naked hand		X		X				
Sapajus nigritus	Mico		X		X				
Leopardus wiedii	Maracaja Cat		X		X				
Cuniculus paca	Paca		X		X				

Caption: AR: Trap; EN: Interview; VE: Traces; RB: Bibliographic References; VD: Direct Visualisation; FE: Faeces; PE: Footprints; VU: Vulnerable; CR: Critically Endangered and EM: Endangered.

Source: The author, 2017.

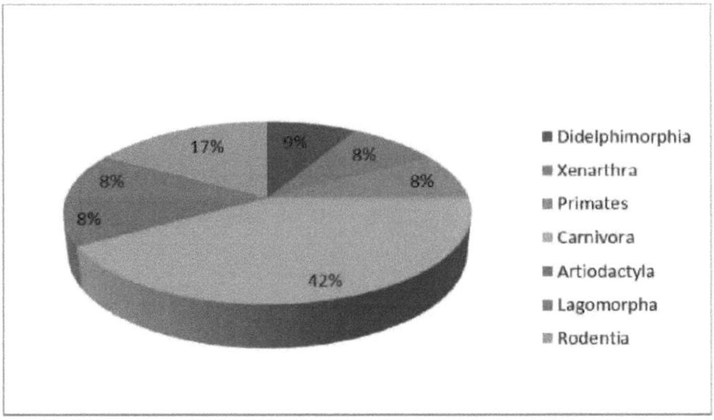

Graph 1 - Order of species found in the study area.
Source: The author, 2017.

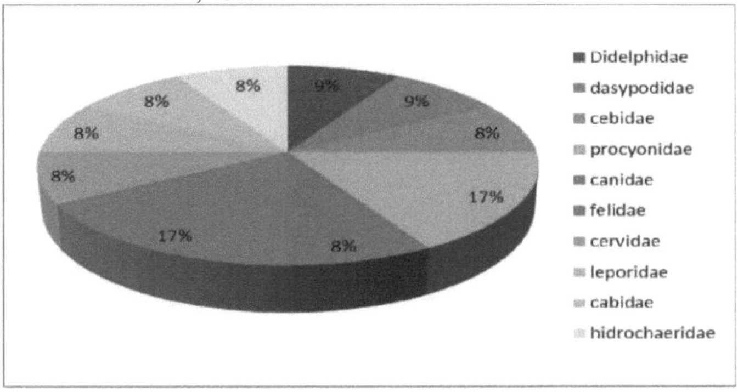

Graph 2 - Families of species found in the study area.
Source: The author, 2017.

In a study on the fauna of the western region of Santa Catarina carried out by Cherem (2008) for the construction of the Quebra Queixo HPP in the municipality of São Domingos, the author mentions that 45 species of native mammals were recorded, the vast majority of which were small. In the work carried out, 12 species of mammals were recorded, or 26.7% of the result presented by Cherem (2008).

There were similarities between the species found in the study region compared to Cherem's work (2008), namely the Mulita armadillo *(Dasypus hybridus),* Catingueiro deer *(Mazama gouazobird),* Quati *(Nasua nasuã),* White-eared opossum *(Didelphis albiventris),* Capybara *(Hydrochaeris hydrochaeris),* Graxaim or Cachorro-do-mato *(Cerdocyon thous)* and Mico *(Sapajus nigritus).*

Cherem (2008) also states that factors such as the alteration and fragmentation of natural environments, hunting and the introduction of exotic species have contributed to the extinction of mammals in the study area and in the region. Large mammals such as the jaguar *(Panthera onca)* and the cateto deer *(Pecari tajacu)* reflect a pattern known as ecological truncation, i.e. a major loss of biodiversity due to the fact that species occur in very low densities.

In the Simplified Environmental Study (EAS) carried out by Construtora Gomes Lourenço Ltda (2006) for the implementation of the Santa Luzia SHPP located on the border between the municipalities of São Domingos and Ipuaçu, 46 species of native mammals were recorded. In the work carried out, 12 species of mammals were recorded.

4.2 FLORA OF THE STUDY REGION

4.2.1 Use of Sampling and On-site Inventory

Analysis of the phytosociological data of the vegetation in the study area of the Ludesa SHPP, located in the municipality of São Domingos - SC, showed that 381 individuals were recorded, distributed among 27 species and 17 different botanical families. The most representative species were *Nectandra lanceolata* with 19.95%, followed by *Cupania vemalis* with 11.81%, *Allophylus edulis* with 9.97% and *Allophylus guaraniticus* with 9.71% of the species sampled, as can be seen in Graph 03.

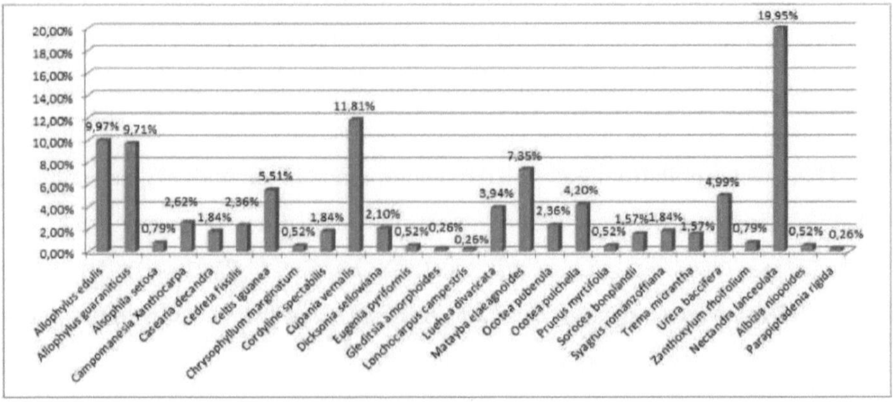

Graph 3 - Species found in the study area
Source: The author, 2017.

The most representative families are Sapindaceae with 38.85 per cent and Lauraceae with 26.51 per cent of the species found. The species from the Sapindaceae family found are *Matayba elaeagnoides, Cupania vemalis, Allophylus edulis* and *Allophylus guaraniticus*. The species from the Lauraceae family found are *Nectandra lanceolata, Ocotea puberula and Ocotea pulchella*, as can be seen in Graph 04.

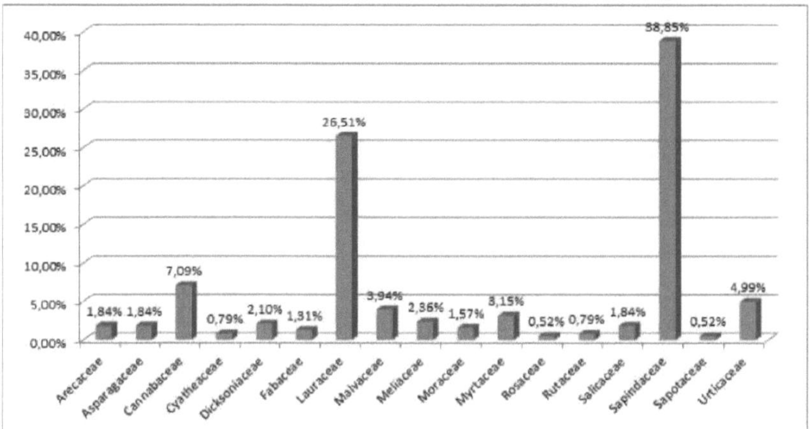

Graph 4 - Families found in the study area.

Source: The author, 2017.

Similar results were found by Santos (2012) in his floristic and phytosociological survey of a forest fragment in the municipality of Faxinai dos Guedes, SC. The most abundant families found were Lauraceae, Mrytaceae followed by Fabaceae, Sapindaceae and Bignoniaceae; the latter family did not have any individuals in the São Domingos study area fragment.

In the study area, there are individuals that, according to MMA decree no. 443 of 17/12/2014, are on the list of endangered species, namely *Dicksonia sellowiana* and *Cedrela fissilis*, which is classified as a vulnerable species.

Cherem (2008) states that the inclusion of species in the Ministry of the Environment's ordinance, as well as their conservation status, is presented by the Botanical Society of Brazil.

Table 1 shows that the species with the highest Absolute and Relative Density were *Nectandra lanceolata* (211.11 trees/ha), *Cupania Vernalis* (125.00 trees/ha),

Allophylus edulis (105.56 trees/ha) and *Allophylus guaraniticus* (102.78 trees/ha).

Comparing the results in Table 1, the most dominant species were *Nectandra lanceolata, Luehea divaricata, Cupania vernalis, Ocotea pulchella* and *Cedrella fissilis.*

Similar results were found by Schom (2014) in his article on the phytosociology of deciduous forest fragments in the state of Santa Catarina - Brazil, where the species that stood out in the tree component due to their high density and dominance values were *Nectandra lanceolata, Luehea divaricata and Cupania vernalis.*

Table 1 - Phytosociological parameters (Density and Dominance) by species for a fragment of Mixed Ombrophilous Forest in the municipality of São Domingos - SC.

Species	ni	DA	DR(%)	DOA	PAIN(%)
Nectandra lanceolata	76	211,11	20	10,53	37
Cupania vemalis	45	125,00	12	2,74	10
Allophylus edulis	38	105,56	10	0,73	3
Allophylus guaraniticus	37	102,78	10	0,66	2
Matayba elaeagnoides	28	77,78	7	2,13	7
Ceitis iguanea	21	58,33	6	0,23	1
Ur era baccifera	19	52,78	5	0,19	1
Ocotea pulchella	16	44,44	4	2,69	9
Luehea divaricata	15	41,67	4	3,90	14
Campomanesia Xanthocarpa	10	27,78	3	0,67	2
Ocotea puberula	9	25,00	2	0,29	1
Cedrela fissilis	9	25,00	2	2,28	8
Dicksonia sellowiana	8	22,22	2	0,19	1
Casearia decandra	7	19,44	2	0,12	0
Syagrus romanzoffiana	7	19,44	2	0,75	3
Cordyline spectabilis	7	19,44	2	0,05	0
Sorocea bonplandii	6	16,67	2	0,10	0
Trema micrantha	6	16,67	2	0,05	0
Zanthoxylum rhoifolium	3	8,33	1	0,05	0
Alsophila setosa	3	8,33	1	0,03	0
Albizia niopoides	2	5,56	1	0,01	0
Prunus myrtifolia	2	5,56	1	0,01	0
Eugenia pyriformis	2	5,56	1	0,03	0
Chrysophyllum marginatum	2	5,56	1	0,27	1
Gleditsia amorphoides	1	2,78	0	0,03	0
Parapiptadenia rigida	1	2,78	0	0,01	0
Lonchocarpus campestris	1	2,78	0	0,01	0

Legend: **Ni** = Total number of trees per plot; **DA** = Absolute Density (trees/ha); **DR** = Relative Density **(%); DOA** = Absolute Dominance (m^2 /ha); **DOR** = Relative Dominance (%).

Source: The author, 2017.

Table 2 shows the Absolute Frequency and Relative Frequency values, the importance value index and the coverage value index. It can be seen that 9 species showed 100% occurrence in the plots, namely *Matayba elaeagnoides, Cupania vemalis, Nectadra lanceolata, Ocotea puberula, Ocotea pulchella, Cedrela fissilis, Allophylus edulis, Ceitis iguanea* and *Allophylus guaraniticus.*

The species with the highest Absolute Frequency values are the same as those with high Relative Frequency values, indicating that the individuals of each species are distributed throughout the area and not grouped together.

Table 2 - Phytosociological parameters (Absolute and Relative Frequency, Importance Value index and Cover) by species for a fragment of Mixed Ombrophilous Forest in the municipality of São Domingos - SC.

Species	ni	FA%	FR%	CVI%	IVI%
Nectandra lanceolata	76	100,00	5,26	28,28	20,61
Cupania vernalis	45	100,00	5,26	10,66	8,86
Luehea divaricata	15	75,00	3,95	8,74	7,14
Matayba elaeagnoides	28	100,00	5,26	7,37	6,67
Ocotea pulchella	16	100,00	5,26	6,78	6,28
Allophylus edulis	38	100,00	5,26	6,26	5,93
Allophylus guaraniticus	37	100,00	5,26	6,01	5,76
Cedrela fissilis	9	100,00	5,26	5,14	5,18
Ceitis iguanea	21	100,00	5,26	3,15	3,86
Ur era baccifera	19	75,00	3,95	2,82	3,20
Campomanesia Xanthocarpa	10	75,00	3,95	2,47	2,96
Ocotea puberula	9	100,00	5,26	1,68	2,88
Syagrus romanzoffiana	7	75,00	3,95	2,22	2,80
Dicksonia sellowiana	8	75,00	3,95	1,39	2,24
Casearia decandra	7	75,00	3,95	1,13	2,07
Cordyline spectabilis	7	75,00	3,95	1,01	1,99
Sorocea bonplandii	6	75,00	3,95	0,97	1,96
Trema micrantha	6	75,00	3,95	0,87	1,90
Zanthoxylum rhoifolium	3	75,00	3,95	0,48	1,64
Chrysophyllum marginatum	2	50,00	2,63	0,73	1,36
Albizia niopoides	2	50,00	2,63	0,28	1,07
Alsophila setosa	3	25,00	1,32	0,45	0,74
Eugenia pyriformis	2	25,00	1,32	0,32	0,65
Prunus myrtifolia	2	25,00	1,32	0,28	0,62
Gleditsia amorphoides	1	25,00	1,32	0,19	0,56
Lonchocarpus campestris	1	25,00	1,32	0,15	0,54
Parapiptadenia rigida	1	25,00	1,32	0,15	0,54

Legend: **AF** = Absolute Frequency (%); **RF** = Relative Frequency (%); **IVI** = Importance Value Index (%); **CVI** = Cover Value Index (%).

Source: The author, 2017.

The species that stood out most according to Cover Value

were *Nectandra lanceolata* (28.28%), *Cupania vernalis* (10.66%) and *Luehea divaricata* with (8.74%). Despite having a large number of individuals, *Urera baccifera* has a relatively low Cover Value index of (2.82%) compared to the other species.

There was similarity between the species with the highest Cover Value in the article on the Phytosociology of fragments of stationary deciduous forest in the state of Santa Catarina - Brazil by Schom (2014), where the author highlights the species *Cupania vemalis* and *Luehea divaricata.*

For the fragment analysed, *Dicksonia sellowiana* had a low relative frequency (3.95%) compared to the other species and to the study carried out by Santos (2012) on the Floristic and Phytosociological Survey of a Forest Fragment in the Municipality of Faxinai dos Guedes, SC where the relative frequency of the species was (7.62%). As it is a native and representative species in the Mixed Ombrophilous Forest, the result shows that the species is gradually disappearing.

In the studies by Vibrans (2012), *Nectandra lanceolata* also appeared among the first in the parameters of dominance and value of importance. In addition to these, others were similar and stood out, such as *Luehea divaricata, Cupania vemalis, Ocotea puberula* and *Cedrela fissilis.* According to Schneider (2014), these species are generalists, benefiting from clearings and common in forest edges.

The species found in this study were cited by Pellizzaro (2009) in his article on the floristic composition and phytosociology of the arboreal component of a mixed ombrophilous forest fragment in the municipality of Clevelândia, Paraná. This indicates that they are not restricted to the study area and the state of Santa Catarina.

According to the species/area curve in Graph 05, the sample sufficiency showed that the number of plots installed was not enough to characterise the vegetation in the area. It can be seen that the curve does not stabilise as the area increases, linked to the fact that the area studied is very large.

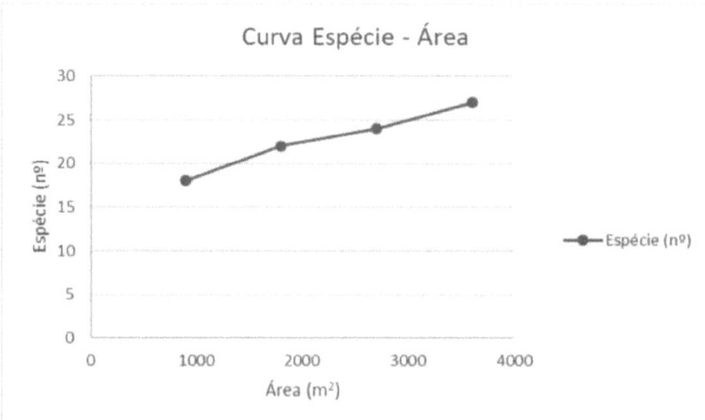

Graph 5 - Species/area curve from the phytosociological study of the area of direct influence of the Ludesa SHPP located in the municipality of São Domingos, Santa Catarina.
Source: The author, 2017.

According to Felfili (2011) floristic sample sufficiency is not only linked to the density and richness of species in an area. The equity of their frequency is also an important factor. In addition, the homogeneity of the area being studied is also relevant. The less variation there is between plots, the quicker stability will be achieved.

4.2.2Use of Satellite Imagery

By analysing the classified images from the different years, it was possible to obtain class data and thus understand the change in the scenario under study, as shown in Table 3.

Table 3 - Land Use Classes

Classes	2004 (ha)	2004 (%)	2017 (ago)	2017 (%)
Forest Cover	116,29	37%	109,66	35%
Agricultural Area	178,71	57%	129,26	41%
Water Resources	19	6%	75,08	24%

Caption: **2004:** Year before the project was installed; area shown in (ha). **2017:** Year after the development was installed; area shown in (ha).
Source: The Author, 2017

It is clear from the satellite images that the legal area of permanent preservation on the banks of the reservoir has increased. The area calculated using the 2004 image, before the reservoir was filled, ranged from 1.0 to 18.0 metres. The 2017 image shows that the APP has increased to 30.0 - 40.0 metres in total area.

Although the current APP has increased in area compared to 2004 (map 01) and 2017 (map 02), it cannot be said that the vegetation cover in the area has also increased. The APP that existed before the reservoir was filled was flooded, and the water level reached places that used to be plantation areas, pastures and even forest cover.

According to the data presented in Table 3, the area directly affected by the development has suffered impacts in all its geographical aspects, such as the removal of vegetation, where there has been a 2% reduction in forest cover, the modification of the topography due to the formation of the lake, resulting in an 18% increase in the water resource and a 16% reduction in the agricultural area.

CHAPTER 5

FINAL CONSIDERATIONS

The implementation of a small dam can generate major environmental impacts as a result of its construction. The purpose of this study was to analyse the impacts of the construction of the Ludesa SHPP in relation to the biotic environment (fauna and flora).

The fauna of the study region has suffered from the impact of construction, where the natural habitat of many species has been diminished and local fauna has been driven away and made susceptible to anthropogenic actions such as hunting and capture. The decrease in fauna was noticeable, with only 12 species of mammals reported in this study, a decrease of 73.3 per cent compared to other studies carried out in the region.

A total of 27 forest species were identified, showing a tendency to maintain the species found, which had the highest value of importance and coverage of the tree component. Even so, a decrease in species diversity and forest structure due to anthropogenic factors was noted.

Faunal and floristic studies in the region are scarce, so the results obtained from this (short-term) study do not fully reflect reality, which leads us to conclude that long-term studies would be necessary to accurately state the results obtained.

Satellite images from 2004 and 2017 show a significant decrease in the area of agricultural use and forest cover due to the filling of the reservoir. However, it can be seen that permanent preservation areas have been improved as a result of the construction of the project.

Although it was observed that the environments have been altered by the construction of the development, several species of conservation interest were found, the main ones for fauna being the Catingueiro deer *(Mazama gouazoubirà) and* flora being the tree *fern (Dicksonia sellowiand),* cedar *(Cedrela fissilis)* and yellow cinnamon *(Nectandra lanceolatd).*

CHAPTER 6

REFERENCES

ABILHOA, V. **Composition and structure of the Ichthyofauna of the Irani Reservoir, Metropolitan Region of Curitiba, Paraná, Brazil.** IV Seminar of the Interdisciplinary Project on Eutrophication of Public Supply Waters in the Altissima Iguaçu Basin. Curitiba - PR, 2003.

ABDO, José. **Resolution N°705, of 17 December 2002.** Federal Official Gazette. December 2002.

BATISTA, M.F Bruna. **Review of the Environmental Impacts generated during the installation phase of hydroelectric dams: An analysis of the upper Juruena-MT sub-basin.** UFMT, Biodiversity V.ll, Nl, p.69, 2012.

BRANCO Jr. et.al. **Identification of mastofauna by traces.** Interciência, 2006.

BRITO D.C Ângela. **Analysing the performance evaluation of the environmental management system (ISO 14001) of hydroelectric power plants: the relationship between the company and the surrounding communities.** Brazilian Journal of Environmental Sciences - Number 24. Macapá - 2012.

BONSOR, Kevin. **Environmental Impacts in the Construction of Hydroelectric Power Plants.** http://ciencia.hsw.uol.com.br/usinas-hidreletricas5.htm Accessed on 22/09/2016.

CARVALHO, Oswaldo. **Footprints: Good practice series.** Belém - PA: Edufra, 2008.

CARVALHO, Diego Lellis. **Methodologies for Evaluating the Environmental Impacts of Hydroelectric Projects.** XVI National Meeting of Geographers, AGB 2010.

CHERREM, Jorge. **The fauna of the areas of influence of the Quebra Queixo Hydroelectric Power Station.** Erechim, RS, 2008.

CRUZEIRO, Aerofoto. **School Atlas of Santa Catarina** / State Secretariat for General Coordination and Planning. Undersecretary of Geography and Statistics. Rio de Janeiro, 1991.

Simplified Environmental Study; **Implementation of the Santa Luzia Small Hydroelectric Power Plant.** July 2006.
FATMA, Instrução Normativa da Produção de Energia Hidroelétrica - IN 44, Fundação do Meio Ambiente, Version March/2012.

FACURI, M.F. **The implementation of hydroelectric power plants and the environmental licensing process: the importance of articulation between the electricity sector and the environment in Brazil.** Master's thesis. 77p. Federal University of Itajubá. Itajubá, MG. 2004.

FILFILI, Jeanini. et.al. **Phytosociology in Brazil - Methods and case studies.** Vol.1 UFV, Viçosa, MG, 2011.

GONSALES. **Amphibians of the Araucarias National Park, Santa Catarina State,** Brazil. Elaine M.L Gonsales, 2009.

IAP. **Guarani River State Park Management Plan. Annex II - Phytosociological Parameters.** Environmental Institute of Paraná, Curitiba, 2002.

KLEIN, R. M. **Aspectos dinâmicos da vegetação do sul do Brasil.** Itajaí, v. 36, n. 36, p. 5-54,1984.

LEPSCH, Igo. **Soil formation and conservation.** IgoF.Lepsch - 2ed. São Paulo, 2010.

LATINI, Juliana. **Examining the Contradictions Surrounding Small Hydroelectric Plants as Sustainable Energy Sources in Brazil.** UFPR - DeMA Development and Environment. VoL, 37 May 2016.

MENDES, Noeli Aparecida. **Hydroelectric plants and their impacts: the socio-environmental and economic aspects of the Rural Resettlement of Rosana - Euclides da Cunha Paulista.** UNESP, Presidente Prudente, 2005.

MMA - **Caderno da Região Hidrográfica do Rio Uruguai -** Ministry of **the** Environment, Secretariat of Water Resources. Brasilia - 2006.

MMA - Environmental Licensing Booklet. National Training Programme for Environmental Managers. Brasilia - 2009.
MORO-RIOS, Rodrigo, et.al. **Manual of tracks of the fauna of Paraná.** IAP - Environmental Institute of Paraná, 2008

NILTON, Cássio Luiz. **The impact of small hydroelectric power stations on the environment.** Postgraduate work, Lavras, Minas Gerais; 2009.

NOBREGA, André. **Request to revoke the authorisation of the Ludesa Small Hydroelectric Power Plant.** May 2012
http://www2.aneel.gov.br/aplicacoes/noticias_area/arquivos/48500.003572.2001.36.pdf Accessed on

16 August 2016.

OLIVEIRA, Guilherme et.al. **Identification of Amazonian mammals through trichological analysis for ecological studies.** XX Jornada de Iniciação Cientifica. PIBIC INPA-CNPq/FAPEAM, Manaus, 2011.

PELLIZZARO, L. **Floristic composition and phytosociology of the tree component of a fragment of mixed ombrophilous forest in the municipality of Clevelândia, Paraná.** Proceedings of the IX Brazilian Ecology Congress, 13 to 17 September 2009, São Lourenço - MG.

REZENDE, Vanessa. **Phytogeographic Patterns of the Tree and Shrub Flora of Santa Catarina.** Postgraduate dissertation in Plant Biology. Institute of Biological Sciences - UFMG, Belo Horizonte, 2013.

RIPPEL, Marcos. **Critical Analysis and Contributions to the Operation and Maintenance of Small Hydroelectric Plants.** Master's thesis. 104p. Institute of Technologies for Development - LATEC, PR, 2013.

SANTOS, C. et.al. **Floristic and phytosociological survey of a forest fragment in the municipality of Faxinai dos Guedes, SC.** Unoesc & Ciência - ACET, v3 nl, p 7-22. Joaçaba jan/jun 2012.

SCHNEIDER, G. et.al. **Floristic and phytosociological survey of the tree component of a fragment of stationary deciduous forest in São Miguel do Oeste, Santa Catarina.** Biotemas, 27(2) pg 43-55. LAVEG, Florianópolis, 2014.

SCHORN, L. et.al. **Phytosociology of deciduous forest fragments in the state of Santa Catarina - Brazil.** Ciência Florestal, Santa Maria v.24, n4, p. 821- 831, Oct-Dec, 2014.

SDR - Secretary of State for Regional Development. Xanxerê: **Regional Characterisation.** 2003.

SILVA. **Diversity of Carnivorous Mammals in Bacaba Park, Nova Xavantina.** São Lourenço, 2011.

SILVA, R.A Diogo. **Basic Hydro-geological Information for the State of Santa Catarina.** XIX Brazilian Symposium on Water Resources. Maceió, 2011.

VIBRANS, Alexandre. **Santa Catarina Forest Inventory: what you should know about Santa Catarina's forests.** Alexandre C. Vibrans - Blumenau, FURB, 2013.

VIBRANS, A. et. al. **Final considerations on the Stationary Deciduous Forest in Santa**

Catarina. Vol. 2. Chap. 10. Inventário florístico Florestal de Santa Catarina. Blumenau: Universidade Regional de Blumenau, 2012c. p. 229-232.

WELTER, Luís. **The geographical space of western Santa Catarina and its environmental cartography.** 91 f. 2006. Dissertation (Master's in Human Geography) - University of São Paulo, São Paulo - SP. 2006.

CHAPTER 7

APPENDIX

A - Photos of the Fauna Survey

FOOTPRINT TRAP

INTERCEPTION AND FALL TRAP

ACTIVE SEARCH

B - Photos of the Flora Survey

FOREST INVENTORY PLOTS

MIX
Papier aus verantwortungsvollen Quellen
Paper from responsible sources
FSC® C105338

Printed by Books on Demand GmbH, Norderstedt / Germany